How Does It Grow?

HOW DO MAPLE TREES GROW?

By Kathleen Connors

Gareth Stevens
PUBLISHING

Please visit our website, www.garethstevens.com. For a free color catalog of all our high-quality books, call toll free 1-800-542-2595 or fax 1-877-542-2596.

Library of Congress Cataloging-in-Publication Data
Names: Connors, Kathleen, author.
Title: How do maple trees grow? / Kathleen Connors.
Description: New York : Gareth Stevens Publishing, [2022] | Series: How does it grow? | Includes index.
Identifiers: LCCN 2020011758 | ISBN 9781538263037 (library binding) | ISBN 9781538263013 (paperback) | ISBN 9781538263020 (6 Pack) | ISBN 9781538263044 (ebook)
Subjects: LCSH: Maple–Juvenile literature. | Maple–Life cycles–Juvenile literature.
Classification: LCC SD397.M3 C66 2022 | DDC 664/.132–dc23
LC record available at https://lccn.loc.gov/2020011758

First Edition

Published in 2022 by
Gareth Stevens Publishing
111 East 14th Street, Suite 349
New York, NY 10003

Editor: Kristen Nelson
Designer: Katelyn E. Reynolds

Photo credits: Cover, p. 1 Matheisl/The Image Bank / Getty Images Plus; pp. 5, 7 Apostrophe Productions/ The Image Bank / Getty Images Plus; pp. 9, 24 (seeds) John Warden/ The Image Bank / Getty Images Plus; p. 11 Kurt Mobus/Getty Images; pp. 13, 24 (seedling) Kerrick/ iStock / Getty Images Plus; pp. 15, 24 (trunk) emer1940/ iStock / Getty Images Plus; p. 17 Anna_Pakutina/ iStock / Getty Images Plus; p. 19 Mark Turner/ Photolibrary / Getty Images Plus; p. 21 Studio Light and Shade/ iStock / Getty Images Plus; p. 23 buzzanimation/ iStock / Getty Images Plus.

Printed in the United States of America

CPSIA compliance information: Batch #CSGS22: For further information contact Gareth Stevens, New York, New York at 1-800-542-2595.

Contents

Maple trees are pretty!
How do they grow?

They can grow many places. It can be warm or cold.

They come from seeds.
The seeds have
little wings.

The seeds grow in soil.
They need water
and sun.

Roots and leaves
start to grow.
These are seedlings.

A stem grows.
It is woody.
This is the trunk.

It grows branches and leaves.

Leaves can be green
or red.
They change color
in fall.

Maple trees make sap.
It is sticky.

Sugar maples have sweet sap.
It can become candy!

Words to Know

seedling

seeds

trunk

Index